BIOCHEMISTRY

Lab Manual

Dr. Shahidul M. Islam

Associate Professor of Chemistry
Delaware State University

innovativeink PUBLISHING
A Division of Kendall Hunt

Cover image Shutterstock.com

innovativeink PUBLISHING
A Division of Kendall Hunt

www.innovativeinkpublishing.com
Send all inquiries to:
4050 Westmark Drive
Dubuque, IA 52004-1840

Print ISBN: 979-8-3197-2097-9
eBook ISBN: 979-8-3197-2098-6

Published in the United States of America

Special thanks to Mehedi Hasan for his contributions with editing, reviewing, and proof reading.

CONTENTS

ACKNOWLEDGMENTS

I would like to express my sincere gratitude to my students, whose curiosity, dedication, and enthusiasm have been a constant source of inspiration throughout the writing of this book. Their thoughtful questions, active engagement, and commitment to learning have significantly enriched the ideas and perspectives presented here.

I would also like to thank my graduate student, Md. Mehedi Hasan, who spent time to edit and proofread this book.

I am also deeply thankful to my family for their unwavering support, patience, and encouragement. Their understanding during the many long hours of writing, along with their constant motivation, made this work possible. This book would not have been completed without their love and support.

I dedicate this book to the loving memory of my late father, Dr. Nurul Islam, whose guidance, values, and inspiration have profoundly shaped my academic journey and continue to influence my work.

Preparation of Solutions of Known Concentration and Related Calculations

1

Introduction

In chemistry and biochemistry, experiments often require the preparation of solutions with precisely known concentrations. A solution is a homogeneous mixture of two or more substances. The substance that gets dissolved is called the **solute** (for example, sodium chloride or glucose), while the medium in which it is dissolved is called the **solvent** (commonly water in laboratory practice). Together, they form a uniform system known as a **solution**. The concentration of a solution states the amount of solute present in a given quantity of solvent or solution.

One of the most widely used concentration units is **Molarity** (M), defined as the number of moles of solute per liter of solution:

$$M = \frac{\text{moles of solute}}{\text{volume of solution (L)}}$$

the number of moles of solute (n) is defined as,

$$n = \frac{\text{mass of solute (g)}}{\text{molar mass }(\frac{g}{mol})}$$

Accurate preparation of solutions with known concentration is essential for reliable experimental results. Incorrect concentrations can lead to errors in chemical reactions, affect reproducibility of results, and compromise biochemical assays. In laboratory practice, even a small deviation in concentration may alter the outcome of titrations, enzyme kinetics experiments, or drug-binding assays.

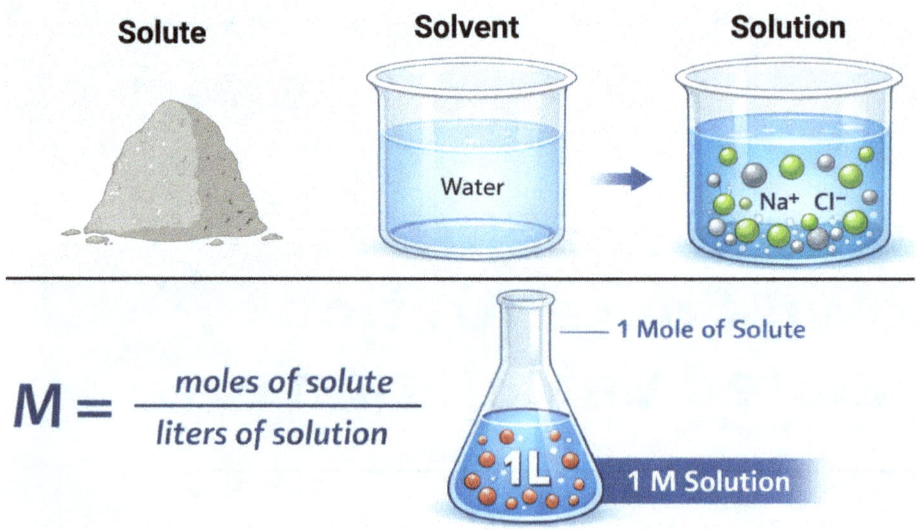

Figure: Illustration of solution formation, concentration, and electrical conductivity. (A) Upper panel: Dissolution of a solute (e.g., NaCl) in water forming a homogeneous solution as ions disperse in the solvent. (B) Lower panel: Concept of molarity showing that solution concentration is defined as moles of solute per liter of solution (M = moles/L).

Beyond molarity, there are other ways to describe the concentration and behavior of solutions. One of these is **Normality (N),** which tells us how many equivalents of a substance are present per liter of solution. For example, sulfuric acid (H_2SO_4) can donate two hydrogen ions (H^+) per molecule, so its normality is twice its molarity in acid-base reactions.

Another important property is **conductivity,** which tells us how well a solution can carry an electric current. This depends on the presence of free-moving ions. Ionic compounds dissolving in water can generate opposite ions that can act as good conductors of electricity. These solutions are also known as **electrolytes.** On the other hand, molecular compounds like glucose dissolve as intact molecules without producing ions. Such solutions are called **nonelectrolytes,** and they do not conduct electricity.

Objectives of this Experiment

The objective of this experiment is to practice calculations and problem-solving related to the preparation and analysis of solutions of known concentration.

1. Calculating the amount of solute required to prepare solutions of defined molarity.
2. Determining the molarity of solutions from given mass and volume data.
3. Convert between common concentration units and apply appropriate unit conversions.
4. Understand the relationship between solute type and the electrical conductivity of the solution.

Concept Review

A. Preparing Solutions (Molarity)

Key Formula:

$$M = \frac{moles\ of\ solute}{volume\ (L)}$$

$$moles = \frac{mass\ (g)}{molar\ mass\ (g/mol)}$$

Example 1: How many grams of NaCl are needed to prepare 500 mL of a 0.50 M solution?

Solution:

$$moles = M \times V = 0.50\ L \times 0.500\ mol/L = 0.25\ mol$$
$$mass = moles \times molar\ mass = 0.25\ mol \times 58.44\ g/mol = 14.61\ g$$

Explanation: Convert volume to liters, then calculate moles and convert to grams.

B. Finding Molarity

Key Formula:

$$M = \frac{mass\ (g)}{molar\ mass\ (g/mol)\ \times\ volume\ (L)}$$

Example 2: What is the molarity of 5.85 g NaCl in 1.0 L?

Solution:

$$moles = 5.85\ g\ \times 1 mol\ /\ 58.44 g$$

Post Lab Quiz

Experiment Title	
Date	
Name	
Lab Partner (s)	

1. **Making NaCl Solution:**

To make 100ml of 0.5 M NaCl solution, how many grams of NaCl do you need if you have 100ml solution?

Show the Calculations below:

Mass of NaCl: ____ (write unit)

Total volume of the solution: ____ (write unit)

Moles of NaCl: __________

Molarity of the solution: ______________ (write unit)

2. **Show work.** What is the molarity of an aqueous solution containing 22.5 g of sucrose ($C_{12}H_{22}O_{11}$) in 35.5 mL of solution?

3. **Show work.** Glucose ($C_6H_{12}O_6$, M= 180.16 g/mol) is a simple sugar and a polar, molecular compound. Blood sugar (glucose) levels are considered normal when the concentration of glucose in the blood is 140 mg/dL two hours after eating.

a) The average adult human body has about 5.0 L of blood. How many moles of glucose are present in 5.0 L of blood when the blood sugar (glucose) concentration is 165 mg/dL?

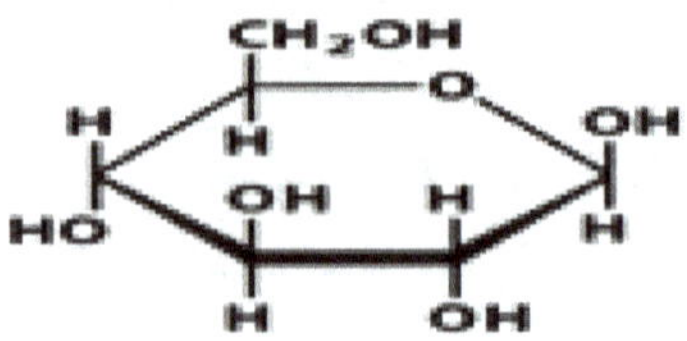

b) What is the molarity of the 165 mg/dL blood glucose solution?

c) i) If glucose was dissolved in water, would this solution conduct an electrical current? Explain your answer.

ii) If NaCl was dissolved in water, would this solution conduct an electrical current? Explain your answer.

Micropipette Measurement Accuracy Test, Dilution of Solutions and Calculating Dilution Factors

2

Introduction

Accurate measurement of small liquid volumes is essential in biochemical experiments. Micropipettes are commonly used to transfer microliter (μL) volumes with high precision. However, their accuracy depends on proper handling and technique.

In this experiment, you will first evaluate the **accuracy and precision of a micropipette** by measuring known volumes of water and comparing them to calculated values based on mass. This helps you understand how reliable your measurements are.

The second part focuses on **solution dilution**, which is the process of reducing the concentration of a solution by adding solvent. The extent of dilution is expressed using the **dilution factor (DF)**:

$$DF = \frac{\text{Final Volume}}{\text{Initial Volume}}$$

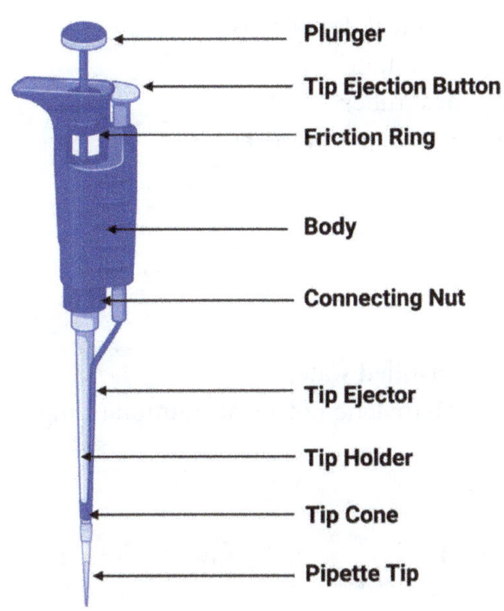

Figure: Different parts of a micropipette

For serial dilutions involving multiple steps, the overall dilution factor is calculated as:

$$DF_{final} = DF_1 \times DF_2 \times DF_3 \times \ldots$$

Understanding dilution is important for preparing solutions of desired concentrations in laboratory experiments.

Objectives of the Experiment

1. To learn proper handling and use of micropipettes
2. To evaluate micropipette accuracy and precision
3. To understand and calculate dilution factors
4. To prepare solutions of known concentrations
5. To analyze concentration differences using absorbance measurements

Required Equipment

1. Micropipettes (2–20 μL, 20–200 μL, 100–1000 μL)
2. Pipette tips
3. Analytical balance
4. Conical flask or vial
5. Volumetric flask (50 mL)
6. Test tubes
7. Spectrophotometer

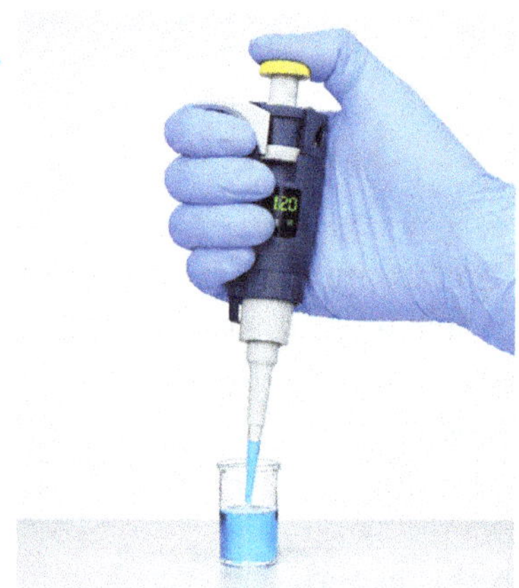

Figure: Dispensing liquid using a micropipette

Required Reagents/Materials

1. Distilled water
2. Methylene red stock solution (1 mg per 100 mL; provided)

Safety Precautions

1. Wear **lab coat, gloves, and goggles** at all times
2. Never pipette by mouth
3. Handle micropipettes carefully (do not invert with liquid)
4. Handle chemical solutions with care (avoid contact)

Waste Disposal

1. Dispose of solutions in **designated liquid waste containers**
2. Discard pipette tips in **appropriate waste bins**
3. Do not pour chemical solutions directly into the sink unless instructed

Experimental Procedure

Part A: Micropipette Accuracy and Linearity Test

A1. Linearity Test

1. Place a clean container on the balance and tare it.
2. Set the micropipette to **200 μL**, dispense water, and record mass.
3. Repeat for **300, 500, 750, and 1000 μL.**
4. Convert mass to volume using:

$$\text{Volume} = \frac{\text{Mass}}{\text{Density}}$$

5. Plot **pipetted volume vs. calculated volume.**
6. Determine **R^2 value** to evaluate linearity.

Record Sheet:

Temperature: **Water density:**

SI	Pipette volume	Measured weight	Calculated Volume (V = mass/density)	R^2 value
1	200 μL			
2	300 μL			
3	500 μL			
4	750 μL			
5	1000 μL			

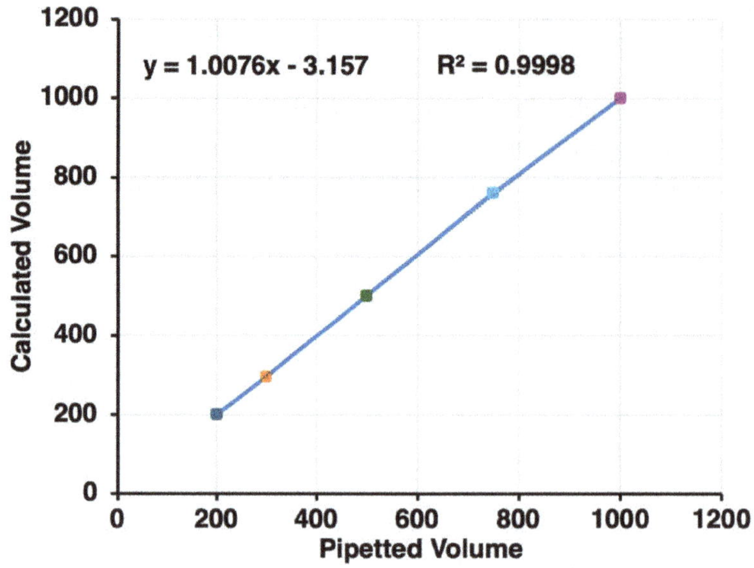

Figure: Scatter plot of pipetted volume *versus* calculated volume for the micropipette linearity test, where the coefficient of determination (R^2) indicates the strength of the linear relationship between expected and measured volumes.

A2. Accuracy (Precision) Test

1. Set micropipette to **250 µL.**
2. Measure and record volume **10 times.**
3. Convert mass to volume for each trial.
4. Calculate **average and standard deviation (SD)**. Lower SD = higher precision

Record Sheet:

Temperature: ***Water density:***

SI	Pipette volume (A)	Measured weight	Calculated Volume (V = weight/density)	Deviation (A-V)
1	250 μL			
2	250 μL			
3	250 μL			
4	250 μL			
5	250 μL			
6	250 μL			
7	250 μL			
8	250 μL			
9	250 μL			
10	250 μL			
Avg	250 μL			
Standard Deviation (SD)				

Part B: Preparation of Diluted Solutions

1. Prepare **Solution A**: 1 mL stock solution + water to 50 mL
2. Prepare **primary dilutions**:
    - Test Tube 1: 1 mL A + 4 mL water
    - Test Tube 2: 0.5 mL A + 2 mL water
    - Test Tube 3: 750 μL A + 3 mL water
3. Prepare **secondary dilutions**:
    - Test Tube 4: 1 mL Tube 1 + 4 mL water
    - Test Tube 5: 1 mL Tube 2 + 4 mL water
    - Test Tube 6: 1 mL Tube 3 + 4 mL water
4. Visually observe color intensity differences
5. Measure absorbance at **520 nm** using water as blank.

Observation / Comments

Post Lab Quiz

Experiment Title	
Date	
Name	
Lab Partner (s)	

1. What do you mean by **dilution factor**?

1. Calculate the **dilution factor** of each of the prepared solution:

   Test tube-4:

   Test tube-5:

   Test tube-6:

3. 10 ml of a stock solution was taken in a 50 ml volumetric flask, and the final volume was adjusted to 50 ml with the solvent. Again 5 ml of this solution was taken in another 100 ml volumetric flask and adjusted the volume to 100 ml by adding the required amount of solvent.

   a) Calculate the dilution factor after making the final diluted solution.

   b) If the concentration of the mother stock solution is given 50 mg/ml, calculate the final concentration of the diluted solution.

Preparing Buffer Solution and Investigating Buffer Capacity

3

Introduction

A **buffer** is a solution that resists changes in pH when small amounts of acid or base are added. It works by neutralizing the added acid or base, keeping the pH of the solution relatively stable. Buffers play a crucial role in various areas of chemistry, particularly when pH control is necessary during reactions, such as in the study of enzymes or proteins. For instance, human blood is buffered to maintain a pH of 7.4, and even slight deviations can lead to serious conditions like acidosis (when the pH drops) or alkalosis (when the pH increases).

To **prepare a buffer solution**, two components are required: one that reacts with OH^- and another that reacts with H^+ or H_3O^+. Buffers are typically made by combining a weak acid with its conjugate base (like acetic acid and sodium acetate) or a weak base with its conjugate acid (like ammonia and ammonium chloride). Generally, buffers are effective within a pH range of ±1 unit around the pKa of the acid. The Henderson–Hasselbalch equation provides the necessary information for preparing buffer solutions.

$$pH = pKa + log\frac{[conjugate\ base]}{[weak\ acid]}$$

There is a limit to the amount of acid or base that can be added to a buffer solution before one of the components is used up. This limit is called the **buffer capacity** and is defined as the moles of acid or base necessary to change the pH of one liter of solution by one unit.

$$Buffer\ Capacity = \frac{[\text{number of moles of } OH^- \text{ or } H^+ \text{ added}]}{pH\ Change\ \times\ Volume\ (L)}$$

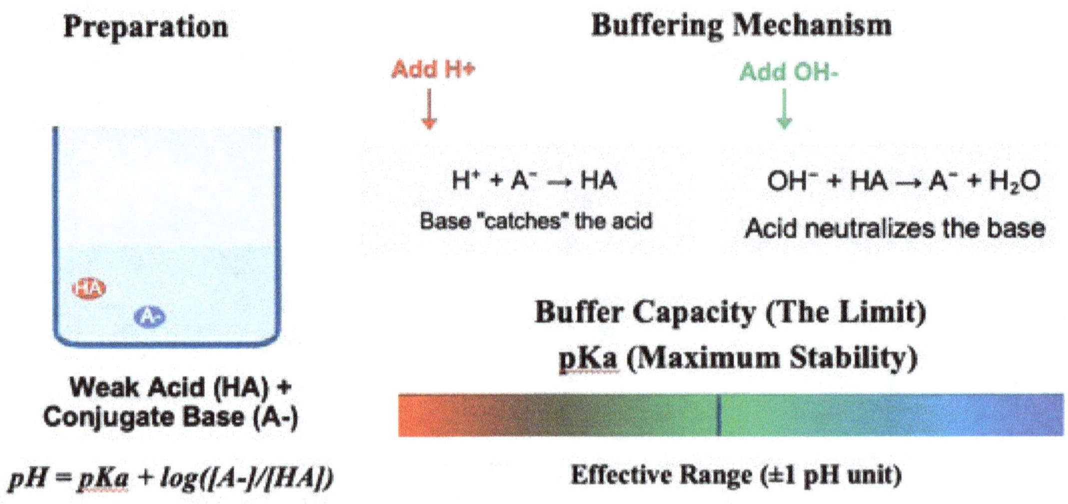

Figure: Illustration of buffer action. A buffer composed of a weak acid (HA) and its conjugate base (A^-) resists pH changes by neutralizing added H^+ or OH^-. Buffering is most effective near the pK_a (±1 pH unit), beyond which the buffering capacity is exhausted.

Objectives of the Experiment

1. To understand the function of buffer solutions
2. To observe how buffers resist pH changes
3. To prepare a buffer solution using weak acid and conjugate base
4. To measure and calculate buffer capacity

Required Equipment

1. Beakers (50–100 mL)
2. Volumetric cylinder / pipette
3. Glass rod
4. pH meter
5. Test tubes

Required Reagents/Materials

1. Standard buffer solution (pH ~7)
2. Distilled water
3. 0.1 M HCl
4. 0.1 M NaOH
5. Universal indicator
6. 0.1 M Acetic acid
7. 0.1 M Sodium acetate

Safety Precautions

1. Wear **lab coat, gloves, and safety goggles**
2. Handle acids and bases carefully (preferably in fume hood)
3. Avoid direct contact with chemicals
4. Report spills immediately

Waste Disposal

1. Dispose of acid/base solutions in **designated liquid waste containers**
2. Do not pour concentrated acids or bases directly into the sink
3. Clean glassware thoroughly after use

Part A: Observing Buffer Action

1. Take two beakers and label them **Water** and **Buffer**.
2. Add 10 mL distilled water to one and 10 mL buffer solution to the other.
3. Add 2 drops of universal indicator to both.
4. Add a few drops of **0.1 M NaOH** to water and observe color change.
5. Add the same number of drops to the buffer solution and observe.
6. Repeat the same steps using **0.1 M HCl**.
7. Record your observations

Observation:

For Water Only:

1. Color change observed for water after adding 0.1 M NaOH : ....

   Amount of 0.1 M NaOH required : .... (Drops)

2. Color change observed for water after adding 0.1 M HCl : ....

   Amount of 0.1 M HCl required : .... (Drops)

For Buffer Solution: (Circle your observation)

1. After adding 0.1 M NaOH: A) Color Changed B) Not Changed
2. After Additing 0.1 M HCl: A) Color Changed B) Not Changed

Part B: Preparation of Acetic Acid Buffer

1. Take a clean beaker.
2. Add **10 mL of 0.1 M acetic acid.**
3. Add **10 mL of 0.1 M sodium acetate.**
4. Mix thoroughly.
5. Measure the pH using a pH meter. [Expected pH ≈ pKa of acetic acid (4.76)]
6. Record your observations

Part C: Measuring Buffer Capacity

1. Take **10–20 mL** of prepared buffer.
2. Measure and record initial pH.
3. Add **0.5 mL of 0.1 M NaOH**, mix, and record pH.
4. Repeat additions until noticeable pH change (0.2–0.5 units). Record the pH values for every steps.
5. Record total volume of NaOH added.

Calculation

Calculate the pH change you observed = Final reading – Initial reading

=

=

The volume of 0.1 M NaOH you added = 0.5 ml + 0.5 ml + 0.5 ml ....

$$Moles\ of\ NaOH = \frac{0.1\ \times\ NaOH\ volume\ added\ in\ ml}{pH\ Change\ \times\ Volume\ (L)}$$

=

Your buffer solution (10/20 ml) in litter = (..........ml)/1000 = ...

$$Buffer\ Capacity = \frac{[\text{number of moles of } OH^{-} \text{ added}]}{pH\ Change\ \times\ Volume\ (L)}$$

=

Observation

Buffer Capacity observed =

Post Lab Quiz

Experiment Title	
Date	
Name	
Lab Partner (s)	

1. What do you mean by **buffer solution**? Explain the importance of buffer in our body.

2. What is an **indicator**?

3. What is the **pH range** for Blood?

4. Write the reaction of acetic acid-sodium acetate buffer solution:

   a) What happens when acidic solution is added to this buffer?

   b) What happens when basic solution is added to this buffer?

4

Isolation and Identification of Protein (Casein) from Milk

Introduction

Casein is the main protein found in milk and is important for nutrition. It can be separated from milk by changing the pH of the solution

Every protein has a special pH called the **isoelectric point**. At this pH, the protein has no overall charge (the positive and negative charges balance each other). When this happens, the protein becomes least soluble in water and starts to stick together, forming solid particles that settle out of the solution (precipitate).

At low pH (acidic conditions), proteins carry a positive charge, and at high pH (basic conditions), they carry a negative charge. In both cases, the charges repel each other, keeping the proteins dissolved in water.

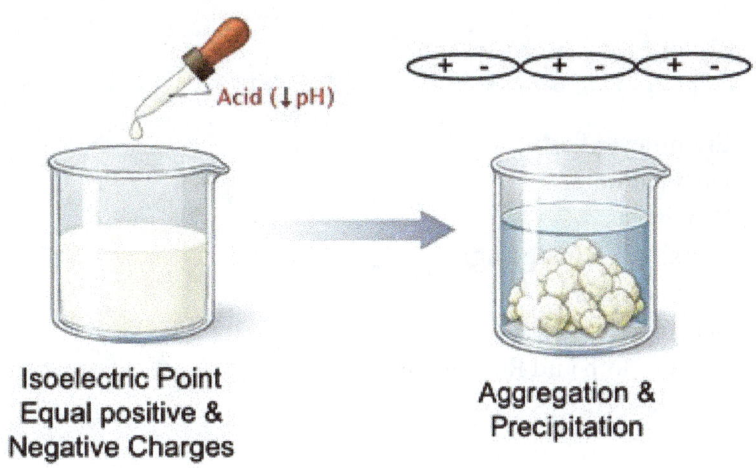

Figure: Proteins tend to aggregate by electrostatic interaction in isoelectric point as the positive end of one protein molecule attracts the negative end of another protein molecule, and the aggregates precipitate out of solution.

In this experiment, milk (which has a pH of about 7) will be acidified to pH 4.6, the isoelectric point of casein. At this point, casein loses its charge and forms a solid precipitate, allowing it to be separated from the liquid.

After isolating casein from milk, the next step is to confirm that the precipitate is actually a protein. This can be done using simple chemical tests that detect characteristic properties of proteins.

One common test is the **Biuret test**. In this test, the sample is treated with copper(II) sulfate. If proteins are present, copper ions interact with peptide bonds and produce a violet (purple) color. This color change confirms the presence of protein.

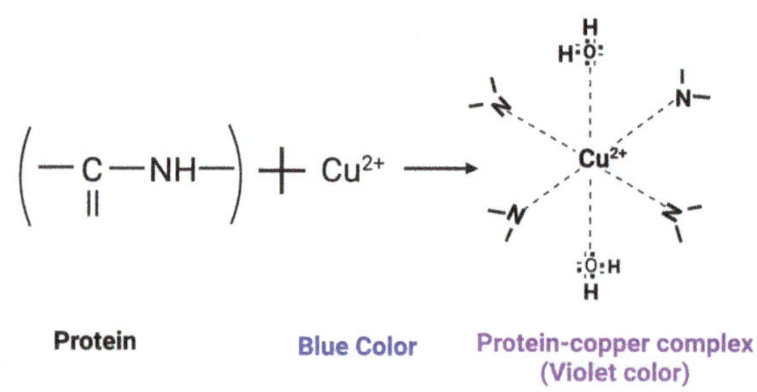

Another test is the **heavy metal test**. When proteins are treated with heavy metal ions (such as lead or mercury salts), the metal ions bind to functional groups in the protein, causing it to denature and form an insoluble precipitate. The formation of this precipitate indicates the presence of protein.

The **coagulation test** is also used to identify proteins. When a protein solution is heated, the protein structure unfolds (denatures) and the molecules aggregate, forming a visible solid or cloudy precipitate. This process is similar to how egg whites turn solid when cooked. The appearance of coagulation upon heating confirms the presence of protein.

Objective of the experiment

1. To isolate casein from milk using pH adjustment
2. To understand the concept of isoelectric precipitation
3. To perform chemical tests to confirm the presence of protein
4. To calculate the percentage of casein in milk

Required Equipment

1. Erlenmeyer flask
2. Distilled water
3. Watch glass
4. Glass rod (For stirring)
5. Thermometer
6. Beaker
7. Cheese cloth and Rubber band
8. Spatula

Required Materials/Reagents

1. Dry milk powder
2. Distilled water
3. Glacial acetic acid
4. 10% NaOH solution
5. 2.5% $CuSO_4$ solution
6. $Pb(NO_3)_2$ solution

Safety Precautions

1. Wear **lab coat, gloves, and goggles**
2. Handle acids and bases carefully
3. Avoid contact with heavy metal solutions
4. Do not overheat the sample

Waste Disposal

1. Dispose of liquid waste in **designated chemical waste containers**
2. Collect heavy metal waste separately (do not pour into sink)
3. Discard solid residues in appropriate waste bins

Experimental Procedure

PART A: Isolation of Casein

1. Weigh **5 g of milk powder (W_0)** into a flask.
2. Add **50 mL distilled water** and mix thoroughly.
3. Heat in a water bath (~40°C) while stirring until dissolved.
4. Add **glacial acetic acid dropwise** with stirring until precipitation forms.
5. Continue adding acid until casein fully precipitates.
6. Filter the mixture using **cheese cloth.**
7. Squeeze gently to remove excess liquid.

Figure: Precipitation of the casein and filtering using cheesecloth

8. Transfer and dry the precipitate. Weigh:
   Empty watch glass (W_1)
   Watch glass + casein (W_2)

Calculation

Weight of milk powder (W_0) = ... g

Weight of watch glass (W_1) = ... g

Weight of Casein with the watch glass (W_2) = ... g

$$\% \, Casein = \frac{Weight \; of \; solid \; casein}{Weigh \; of \; milk \; powder} \times 100 \, \%$$

$$= \frac{W2 - W1}{W0} x \; 100 \, \%$$

=

=

Observation

Before adding glacial acetic acid =

After adding glacial acetic acid =

Casein percentage in the milk sample =

PART B: Identification of Protein

i) Biuret Test

1. Take 1 mL sample solution
2. Add 2 drops of $CuSO_4$
3. Observe color change
4. Violet color confirms protein's presence.

Observation / Comments

ii) Heavy Metal Test

1. Take 1 mL sample solution
2. Add $Pb(NO_3)_2$ solution
3. Observe formation of precipitate
4. White precipitate confirms protein denaturation

Observation / Comments

iii) Coagulation Test (Heat Test)

1. Take 3 mL sample solution
2. Heat in water bath (~3 minutes)
3. Observe changes (Formation of coagulated/solid protein should be observed)

Observation / Comments

Post Lab Quiz

Experiment Title	
Date	
Name	
Lab Partner (s)	

1. The picture below illustrates a hypothetical structure of casein at pH = 7.4, showing the charge on several sidechains.

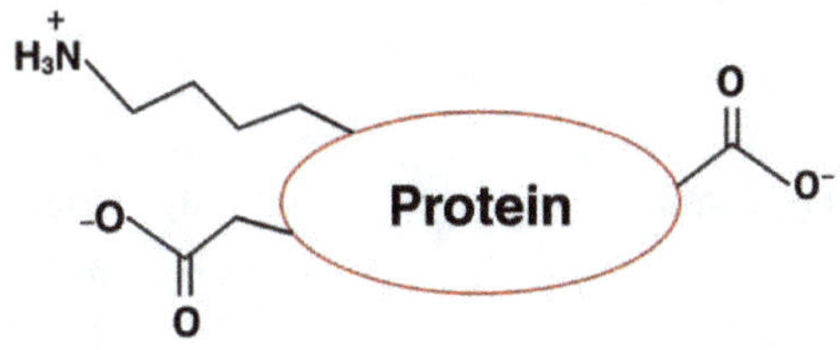

i) Draw a molecular-level picture of the same protein model at pH = 3.0. Indicate the overall charge of the protein.

ii) Draw a molecular-level picture of the same protein model at its isoelectric point. Indicate the overall charge of the protein.

2. Casein has an isoelectric point at pH 4.6. What kind of charges will be on the casein in its native environment, that is, in milk?

3. Draw the structure of a peptide bond in a typical protein molecule.

4. Explain why casein precipitates when acetic acid is added to it.

Determination of the Concentration of Unknown Protein Sample using Bradford Assay

5

Introduction

Quantifying protein concentration is a fundamental step in biochemical and molecular biology experiments, as it directly impacts downstream analyses such as enzyme assays, electrophoresis, and binding studies. Among the available methods, the Bradford assay is a widely adopted colorimetric technique due to its speed, simplicity, and sensitivity.

The assay relies on the interaction between proteins and Coomassie Brilliant Blue G-250 dye. In an acidic environment, the dye exists in a reddish-brown form; however, upon binding primarily to basic (arginine, lysine, histidine) and aromatic amino acid residues, it undergoes a structural transition to a stable blue form. This shift is accompanied by an increase in absorbance at 595 nm, which can be quantitatively measured using a spectrophotometer.

The intensity of the blue color is directly proportional to the protein concentration within a certain range. By preparing a calibration (standard) curve using a known protein standard; commonly bovine serum albumin (BSA), the concentration of an unknown protein sample can be accurately determined.

The Bradford assay is particularly advantageous for routine laboratory use because it requires minimal sample preparation, provides rapid results, and is sensitive to low protein concentrations. In this experiment, the assay will be used to determine the concentration of an unknown protein sample through spectrophotometric measurement and comparison with a standard curve.

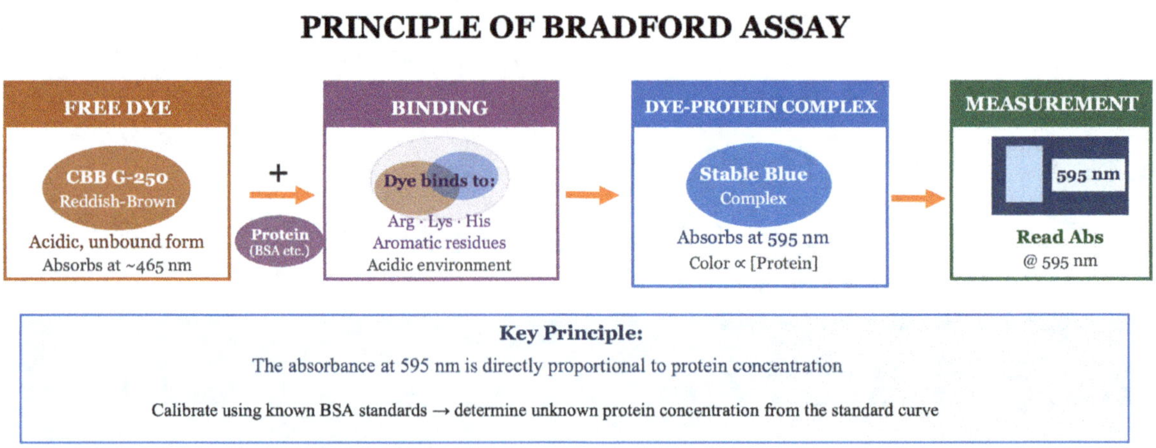

Figure: Principle of Bradford assay

Objectives of the Experiment

1. To understand the principle of the Bradford assay
2. To prepare a standard calibration curve using BSA
3. To measure protein concentration using spectrophotometry
4. To determine the concentration of an unknown protein sample

Required Equipment

1. Test tubes
2. Micropipettes and tips
3. Spectrophotometer
4. Cuvettes

Required Reagents/Materials

1. Coomassie Brilliant Blue G-250
2. Ethanol (95%)
3. Phosphoric acid (85%)
4. Distilled water
5. Bovine Serum Albumin (BSA)
6. Unknown protein sample

Safety Precautions

1. Wear **lab coat, gloves, and goggles**
2. Handle acids and dye solutions carefully
3. Avoid contact with skin and eyes
4. Dispose of chemical solutions properly

Waste Disposal

1. Dispose of dye-containing solutions in **chemical waste containers**
2. Do not pour Bradford reagent into the sink
3. Discard used tips and tubes in proper waste bins

Experimental Procedure

A. Preparation of Bradford Reagent

1. Dissolve **10 mg Coomassie dye** in 5 mL ethanol
2. Add 10 mL phosphoric acid and 85 mL distilled water
3. Mix well and store in a dark bottle

B. Preparation of Standard Protein Solutions

1. Prepare **1 mg/mL BSA stock solution**
2. Label 5 test tubes (Std-1 to Std-5)
3. Add BSA stock to obtain concentrations:
   - 0.2, 0.4, 0.6, 0.8, 1.0 mg/mL
4. Make final volume **5 mL** using distilled water

C. Dilution and Reaction Setup

5. Take new tubes and add **950 µL distilled water**
6. Add **50 µL of each standard solution**
7. Add **3 mL Bradford reagent** to each tube
8. Mix and incubate for **5 minutes**

QUICK REFERENCE TABLE

Bradford Protein Assay | BSA Stock: 1 mg/mL | Measurement wavelength: 595 nm

Tube	BSA Stock (mL)	H_2O added (mL)	Total Vol (mL)	[BSA] mg/mL	Bradford Reagent (mL)
Std-1	1.0	4.0	5.0	0.2	3.0
Std-2	2.0	3.0	5.0	0.4	3.0
Std-3	3.0	2.0	5.0	0.6	3.0
Std-4	4.0	1.0	5.0	0.8	3.0
Std-5	5.0	0.0	5.0	1.0	3.0
S1, S2, S3	50 µL unknown	950 µL	1.0 mL (diluted)	Unknown (?)	3.0

Dilution Step (for Std-1.2 to Std-5.2):
Transfer 50 µL from each Std → add 950 µL dH_2O

⏱ **Timing Critical:**
Measure absorbance at 595 nm within 5–10 min

D. Preparation of Unknown Sample

9. Dilute unknown sample as follows: 50 µL sample + 950 µL water
10. Add **3 mL Bradford reagent**
11. Prepare in **triplicate (S1, S2, S3)**
12. Incubate for 5 minutes

E. Measurement and Analysis

13. Set spectrophotometer to **595 nm**
14. Use distilled water as blank
15. Measure absorbance of standards and samples
16. Plot **standard curve (Absorbance vs Concentration)**
17. Determine unknown concentration using the equation:

$$Y = mX + C$$

$$\Rightarrow \; X = \frac{Y-C}{m}$$

Where:

Y = absorbance
X = concentration

18. Calculate concentration for S1, S2, S3 and take average

Concentration

*You can use the provided Excel sheet to perform the calculation.

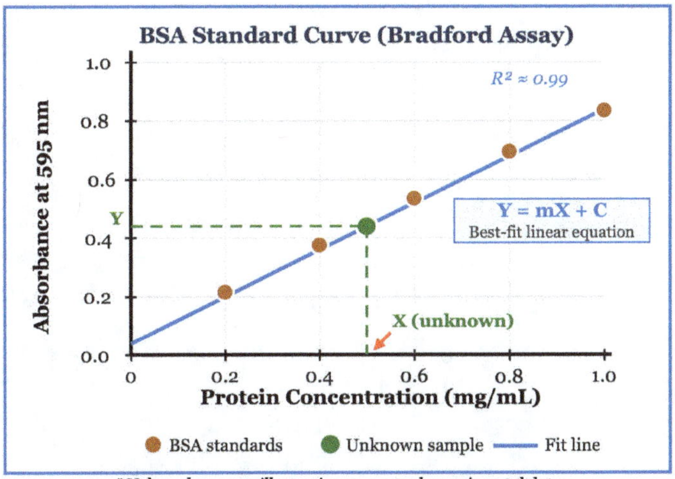

* *Values shown are illustrative; use actual experimental data*

CALCULATION STEPS

Variables

Y = Absorbance (measured at 595 nm)
X = Protein concentration (mg/mL)
m = Slope of standard curve
C = Y-intercept value

Linear Equation:

$$Y = mX + C$$

Solve for Concentration:

$$X = (Y - C) / m$$

For triplicate unknowns (S1, S2, S3):

① Calculate X for each of S1, S2, S3
② Mean = (X_S1 + X_S2 + X_S3) / 3

Observation / Comments

Post Lab Quiz

Experiment Title	
Date	
Name	
Lab Partner (s)	

1. Write the principle of Bradford Protein Assay.

2. What are the advantages of Bradford protein Assay technique?

3. Write the importance and application of protein concentration measurement.

4. Report the concentration of the unknown protein sample from your experiment and comment on your findings.

Enzyme Kinetics Study: K_m and V_{max} Determinations

6

Introduction

Enzymes are biological catalysts that speed up chemical reactions. To understand how efficiently an enzyme works, we study **enzyme kinetics**, which examines how reaction rate changes with substrate concentration.

In this experiment, the substrate **p-nitrophenyl phosphate (PNPP)** is used. When PNPP is hydrolyzed by the enzyme **acid phosphatase**, it produces **p-nitrophenolate**, a yellow-colored product that absorbs light at **405 nm**. This allows enzyme activity to be measured using a spectrophotometer.

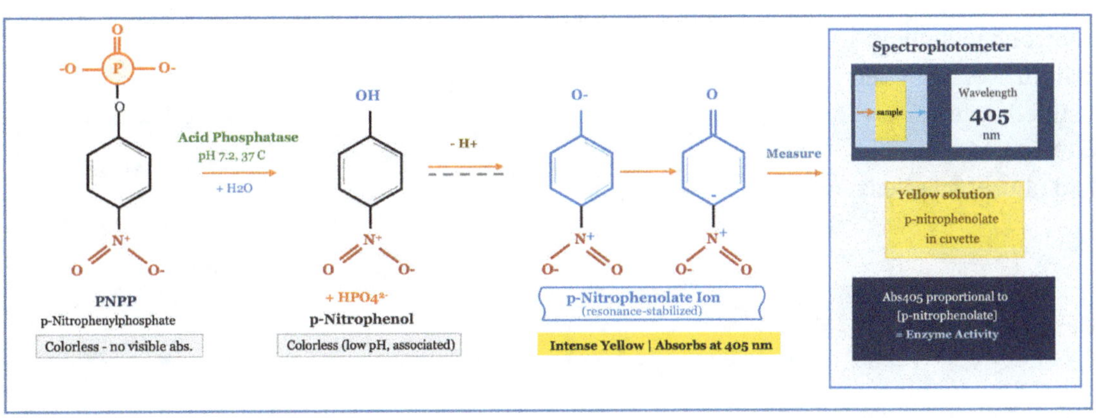

Figure: Principle of the experiment: The Enzyme Kinetics Study

The reaction rate depends on substrate concentration and follows **Michaelis–Menten kinetics**. Two important parameters are:

- **Vmax**: Maximum reaction rate when the enzyme is saturated
- **Km**: Substrate concentration at which the reaction rate is half of Vmax

These values help describe enzyme efficiency and substrate affinity.

Objectives of the Experiment

1. To measure enzyme activity using a colorimetric method
2. To determine reaction rates at different substrate concentrations
3. To calculate Km and Vmax values
4. To understand Michaelis–Menten kinetics

Required Equipment

1. Test tubes
2. Micropipettes and tips
3. Spectrophotometer
4. Stopwatch
5. Glass rod or pipette (for mixing)

Required Reagents/Materials

1. p-Nitrophenyl phosphate (PNPP)
2. Sodium acetate buffer
3. Distilled water
4. NaOH (1 M)
5. Acid phosphatase enzyme

Safety Precautions

1. Wear **lab coat, gloves, and goggles**
2. Handle NaOH carefully (corrosive)
3. Avoid contact with chemicals
4. Use proper pipetting techniques

Waste Disposal

1. Dispose of reaction mixtures in **designated chemical waste containers**
2. Do not pour enzyme or chemical solutions directly into the sink
3. Discard used tips and tubes properly

Experimental Procedure

A. Preparation of Solutions

1. **Preparation of 0.0016 M PNPP stock solution (The molecular weight of PNPP is 371.14g/mol)**
   Weigh 0.119 g of PNPP and dissolved in 200 ml of distilled water.

2. **Preparation of 0.5 M Sodium acetate buffer**
   Dilute 16.67 ml of 3M sodium acetate stock solution to 100 ml with water.

3. **Preparation of 1.0 M Sodium Hydroxide**
   Dissolve 4.0 g of NaOH pellets in 100 ml distilled water.

4. **Preparation of enzyme solution**
   Dissolve 0.125g of wheat germ in 25 ml of distilled water.

B. Preparation of Reaction Mixtures

1. Label four test tubes: A, B, C, D (different substrate concentrations)
2. Add PNPP, buffer, water, and enzyme as per table
3. Mix thoroughly and start timing immediately

Table: Preparation of reaction mixture

Reagents	Reaction Mixtures			
	(800 µl/ml)	(400 µl/ml)	(200 µl/ml)	(100 µl/ml)
PNPP	**1500 µl**	**750 µl**	**375 µl**	**187.5 µl**
Buffer	**0.6 ml**	**0.6 ml**	**0.6 ml**	**0.6 ml**
H_2O	**0.4 ml**	**0.8 ml**	**1.6 ml**	**3.2 ml**
Enzyme	**0.5 ml**	**0.5 ml**	**0.5 ml**	**0.5 ml**

C. Sampling and Reaction Termination

4. **Prepare tubes first (before starting):**
   - Label 8 new test tubes as
     - **A1, B1, C1, D1**
     - **A2, B2, C2, D2**
   - Add **1 mL NaOH** to each of the test tubes.

5. **At 10 minutes:**
   - Take **1 mL** from A, B, C, D to **A1, B1, C1, D1 as follows:**
     - A → A1
     - B → B1
     - C → C1
     - D → D1
   - Shake, reaction stops because NaOH stops the reaction.

6. **At 20 minutes:**
   - Take **1 mL** again from A, B, C, D to **A2, B2, C2, D2 as follows:**
     - A → A2
     - B → B2
     - C → C2
     - D → D2
   - Shake, reaction stops because NaOH stops the reaction.

D. Measurement

7. Measure absorbance at **405 nm** (Note: use water to calibrate at first)
8. Record absorbance for all samples

E. Data Analysis

9. Calculate reaction rate:

$$\text{Rate} = \frac{\text{Absorbance}}{\text{Time}}$$

10. Convert absorbance to concentration using extinction coefficient

11. Plot:
    - **Michaelis–Menten curve** (Velocity vs Substrate concentration)
    - **Lineweaver–Burk plot** (1/V vs 1/[S])

12. Determine:
    - **Vmax** from y-intercept
    - **Km** from slope or x-intercept
    - Use the excel template for calculations (will be provided by the instructor, instructor will make the excel template)

Enzyme Kinetics: Michaelis-Menten & Lineweaver-Burke Analysis

Velocity-Substrate Relationship | Equations | Double-Reciprocal Plot

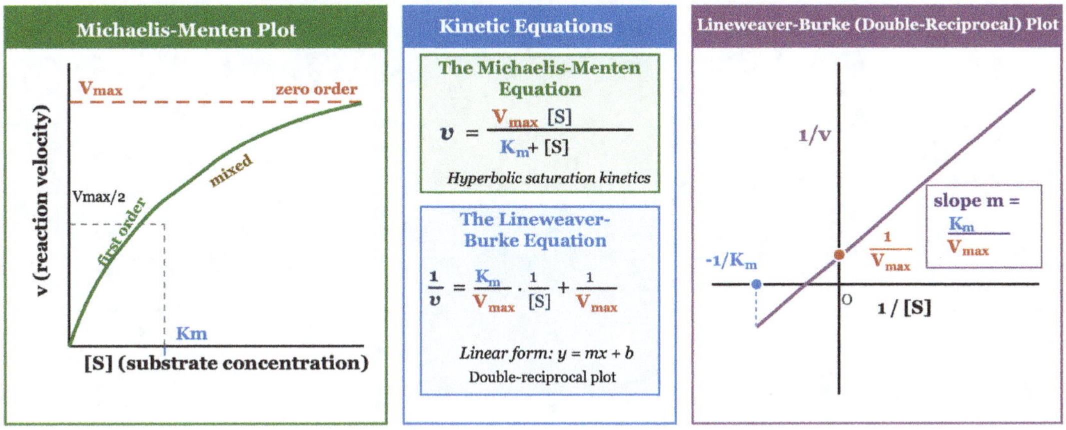

Km = substrate concentration at half-maximum velocity | Vmax = maximum reaction velocity | [S] = substrate concentration

Observation / Comments

Post Lab Quiz

Experiment Title	
Date	
Name	
Lab Partner (s)	

1. Explain how substrate concentration influences enzyme reaction rate. In your answer, describe the differences between first-order and zeroth-order kinetics.

2. Define **Vmax** and **Km**. How can each of these values be determined from a Michaelis–Menten plot?

3. Why is the Lineweaver–Burk plot considered useful in enzyme kinetics? What information can be obtained from its slope and intercepts?

4. In the PNPP assay, why is absorbance measured at **405 nm**? What does an increase in absorbance indicate about the reaction?

5. Why was **pH 7.2** selected for the enzyme assay instead of a higher pH such as 9.0? Explain the trade-off involved.

Determination of pka and pI Values from Amino Acid Titration Curve

7

Introduction

Amino acids are the basic units that make up proteins, and proteins are essential for almost all functions in living cells. Each amino acid contains two important groups: an amino group ($-NH_2$) and a carboxyl group ($-COOH$). Because of these groups, amino acids can act as both acids and bases. This property is called *amphoteric behavior.*

The form of an amino acid in solution depends on the pH. At low pH (acidic conditions), the amino acid tends to carry a positive charge. At high pH (basic conditions), it tends to carry a negative charge. Between these conditions, there is a point where the positive and negative charges balance each other.

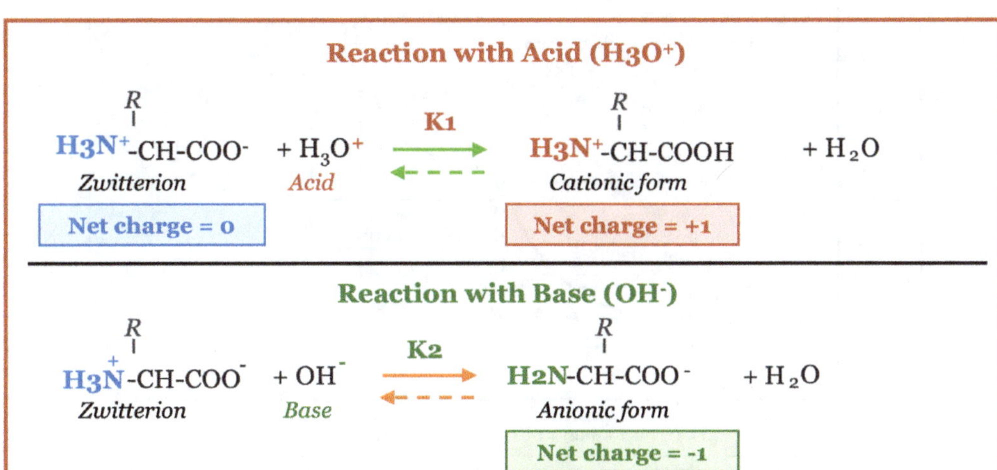

Titration curves help us understand how amino acids behave as the pH changes. For a simple amino acid like glycine, there are two main steps:

- First, the carboxyl group loses a proton (H^+) at low pH
- Second, the amino group loses a proton at higher pH

Each of these steps occurs at a specific pH value called the *pKa*. These values can be found from the flat (buffering) regions of the titration curve.

There is also a special pH where the amino acid has no overall charge. At this point, it exists mainly as a **zwitterion** (a molecule with both positive and negative charges but overall neutral). This pH is called the **isoelectric point (pI).**

For simple amino acids, the isoelectric point can be estimated using:

$$pI = \frac{1}{2}(pK1 + pK2)$$

where:

- $\mathbf{pK_1}$ is for the carboxyl group
- $\mathbf{pK_2}$ is for the amino group

Titration curves are also helpful in the identification of amino acids as follows:

- The number of pKa values differentiates polar and nonpolar amino acids from charged amino acids.
- The position of the pKa values for charged amino acids allows one to identify positively charged from negatively charged amino acids.
- Comparisons between experimental and literature pKa values can allow the identification of a specific amino acid.

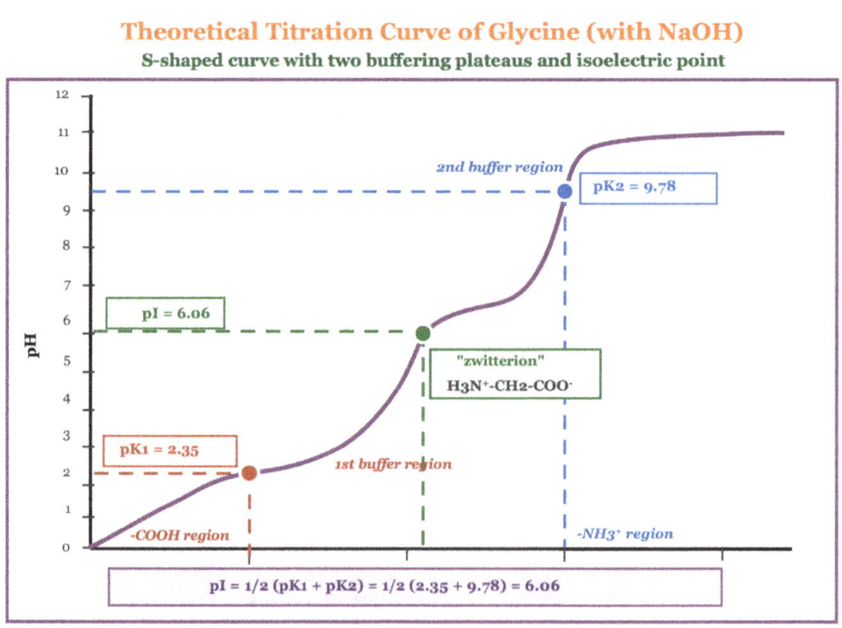

Objectives of the Experiment

1. To construct a titration curve for glycine
2. To determine pKa values from the curve
3. To calculate the isoelectric point (pI)
4. To understand acid–base behavior of amino acids

Required Equipment

1. Beaker or conical flask
2. Micropipette or burette
3. pH meter
4. Glass rod

Required Reagents/Materials

1. Glycine
2. Concentrated HCl
3. 1 M NaOH
4. Distilled water

Safety Precautions

1. Wear **lab coat, gloves, and goggles**
2. Handle HCl and NaOH carefully (corrosive)
3. Avoid direct contact with chemicals
4. Work carefully when measuring pH

Waste Disposal

1. Dispose of acidic and basic solutions in **designated waste containers**
2. Neutralize strong acids/bases if required before disposal
3. Clean glassware properly after use

Experimental Procedure

A. Preparation of Glycine Solution

1. Weigh **3.0 g glycine**
2. Dissolve in ~80 mL distilled water
3. Make final volume to **100 mL**
4. Adjust pH to ~1.5 using HCl

B. Titration

1. Take glycine solution in a beaker
2. Measure and record **initial pH**
3. Add **0.5 mL of 1 M NaOH**
4. Mix well and record pH
5. Repeat addition of NaOH stepwise
6. Continue until **12 mL NaOH** is added
7. Record pH after each addition

Calculation

*Use the provided excel sheet to perform the calculation.

$$pI = \frac{1}{2}(pK_1 + pK_2)$$

Observation / Comments

Post Lab Quiz

Experiment Title	
Date	
Name	
Lab Partner (s)	

1. Define isoelectric point and Zwitterion of amino acids with examples.

2. A titration curve for glutamic acid has three pKa values, two in the acidic pH region, pK1 (α carboxyl group) = 2.2; pK2 (γ carboxyl group) = 4.3; and one in the basic pH region, pK3 (α amino group)= 9.7. Determine the pI of glutamic acid.

3. Show different ionic and dissociation state of amino acids in three different pH conditions (pH = pI, pH < pI, pH > pI).

4. Show the structures of three different types of amino acids

a) Acidic

b) Basic

c) Neutral

Synthesis and Purification of Aspirin

8

Introduction

Aspirin (acetylsalicylic acid) is a widely used medicine that helps reduce pain, fever, and inflammation. It was originally developed from salicylic acid, a compound found in willow bark that has been used for centuries as a natural remedy. However, salicylic acid can irritate the stomach, so it is converted into aspirin, which is safer to use.

In this experiment, aspirin will be prepared by reacting **salicylic acid** with **acetic anhydride** in the presence of a small amount of **acid catalyst (sulfuric acid)**. This reaction forms an **ester**, which is aspirin. Water is then added to stop the reaction and help aspirin form solid crystals.

Aspirin is not very soluble in water, so it precipitates out and can be collected by **filtration**. Other substances remain dissolved in the liquid.

Finally, the purity of the aspirin will test using **iron (III) chloride ($FeCl_3$)**. This reagent reacts with phenol groups to produce a color. Salicylic acid contains a phenol group, but aspirin does not. Therefore, a strong color indicates the presence of unreacted salicylic acid (impurity).

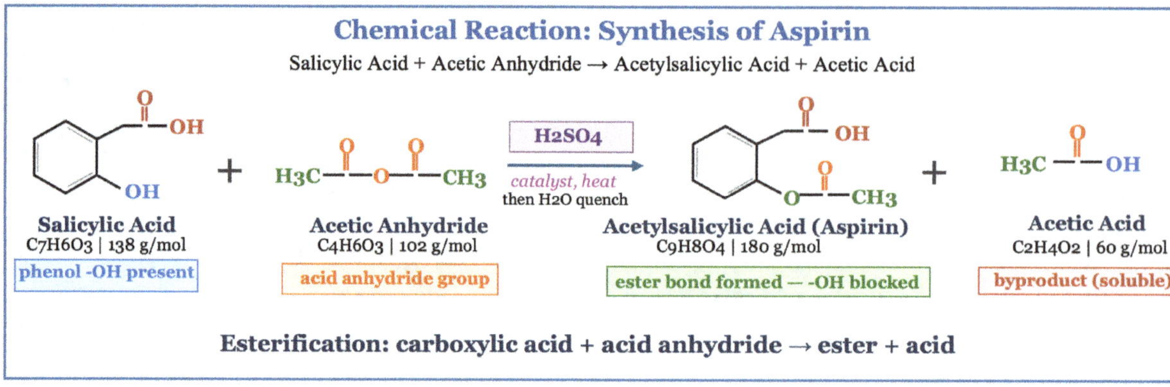

Objectives of the Experiment

1. Chemical Synthesis of Aspirin
2. Purification of Aspirin by recrystallization technique
3. Analysis of the purity of Aspirin

Required Equipment

1. Erlenmeyer flask, Test tube,
2. Hotplate and magnetic stir bar
3. Water bath
4. Thermometer
5. Glass rod
6. Vacuum Filtration setup

Required Reagents/Materials

1. Salicylic acid
2. Acetic anhydride
3. Concentrated H_2SO_4
4. Ice-water
5. Absolute Ethanol
6. 95% Ethanol
7. 2.5% $FeCl_3$

Safety Precautions

1. Use of eye protection is recommended for all experimental procedures (Safety goggles)
2. Hand gloves to avoid any contact of chemicals
3. Lab coat
4. Sulfuric acid (H_2SO_4) is highly corrosive. Avoid contact with your eyes, skin, and clothing. In case of contact, rinse with plenty of water. Ask instructor to assist with the cleaning of any spills.
5. Acetic anhydride is a lachrymator (its vapor irritates the eyes causing tears to flow). Keep it in the fume hood.

Waste Disposal

1. The reaction mixtures used in this experiment may be safely disposed of in the designated container.
2. Solids, aspirin and salicylic acid, should be disposed of in the trash container.
3. Throw away the filter paper in the trash. Dispose of the filtrate (the aqueous solution in the vacuum filtration flask) in the designated containers.

Experimental Procedure:

Procedure for Synthesis of Aspirin:

1. Prepare a **water bath** by filling a 600 mL beaker with ~300 mL water and placing it on a hot plate.
2. Weigh about **2.1 g of salicylic acid**, record the mass, and transfer it to a **125 mL Erlenmeyer flask**. Add a magnetic stirrer in Earleen Myer flask.
3. In a **fume hood**, add **5 mL acetic anhydride**, a **stir bar**, and **~5 drops of sulfuric acid (catalyst)** to the flask.
4. Place the flask in the **water bath** and heat at ~60°C for about **5 minutes**, until the solid dissolves. Remove and allow it to cool.
5. As the solution cools, **crystals will form**. If needed, gently scratch the flask to start crystallization.
6. Add **25 mL cold water**, then place the flask in an **ice bath** to complete crystallization.
7. Collect the crystals by **vacuum filtration**, rinsing the flask to transfer all solids.
8. Wash the crystals with **~15 mL cold water**, then continue filtration to remove liquid.
9. Let the crystals **air dry under vacuum**, then collect them in a beaker.
10. Save a small portion in a test tube labeled **"crude aspirin"** for later analysis.

Purification procedure:

1. Preheat **ethanol (EtOH)** and water on a hot plate. Add a small amount of **EtOH** to the crude aspirin crystals and heat gently (cover with a watch glass).
2. Add more EtOH if needed until the crystals fully dissolve. Once dissolved, add ~1 mL extra EtOH.
3. Add **hot water (about twice the volume of the solution)** slowly.
4. Place the solution in an **ice bath** to allow crystals to form. *(Scratch the glass if needed to initiate crystallization.)* Chill for at least **10 minutes** to maximize crystal formation.
5. Collect the crystals by **vacuum filtration** and wash with a small amount of **cold water**.
6. Dry the crystals. Save a small portion and label it as **"pure aspirin."**

Purity Testing Procedure: checking the presence of any phenolic compound (Salicylic acid in this experiment) in the synthesized aspirin crystal with $FeCl_3$.

Experimental procedure:

1. Take some of salicylic acids in a test tube and label it as salicylic acid.
2. Now we have three test tube for testing (Salicylic acid, crude aspirin and pure aspirin)
3. Add 1 or 2 ml of 95% ethanol to each test tube and dissolve the samples
4. Add one drop 2.5% FeCl3 to each of the test tube.
5. Record your observation.

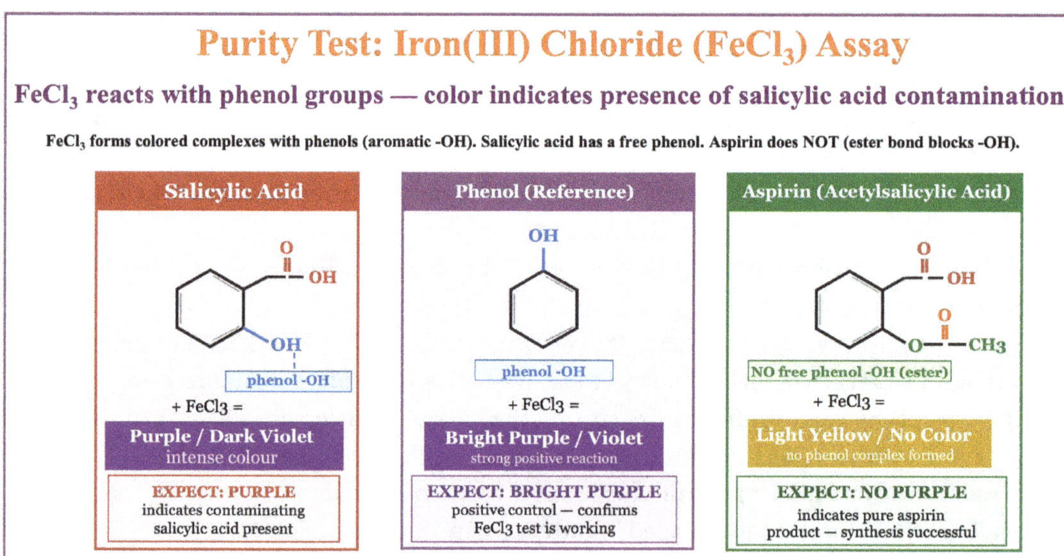

Observation:

Post Lab Quiz

Experiment Title	
Date	
Name	
Lab Partner (s)	

1. What is Aspirin?

2. Write down the chemical reaction equation to synthesize aspirin?

3. Which functional group has changed in the reaction?

4. How much salicylic acid and acetic anhydride were used in your experiment?

5. Why did you use H_2SO_4 and higher temperature?

6. Why did you need to scratch the Erlen Mayer flask?

7. Why did you use ethanol and water to purify the aspirin?

8. What are the color changes you observed after adding $FeCl_3$?

   Salicylic acid: __________

   Aspirin (crude): ___________

   Aspirin (purified): __________

9. Can you guess the color of phenol after reacting with $FeCl_3$, explain your choice.

9

DNA Isolation from Blood Sample

Introduction

DNA (deoxyribonucleic acid) carries the genetic information of all living organisms. In many biological and medical experiments, it is important to extract pure DNA so it can be studied or used in techniques such as PCR, sequencing, and genetic analysis.

Blood is a convenient source of DNA because it contains white blood cells, which have nuclei and therefore contain genomic DNA. Red blood cells do not contain DNA, so they are first removed during the extraction process.

DNA isolation involves a few key steps:

- **Breaking open cells (lysis)** to release DNA
- **Removing proteins and other impurities**
- **Precipitating DNA** so it can be collected

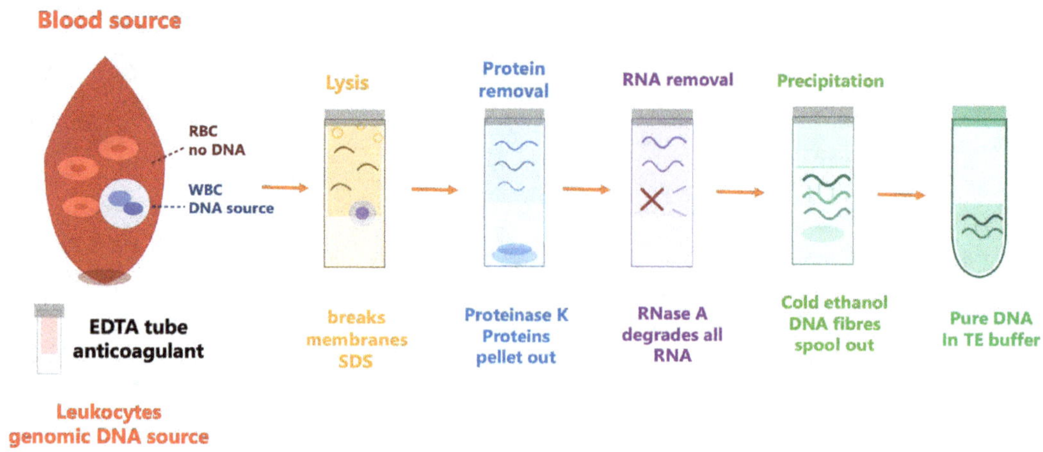

Figure: A schematic of the DNA isolation procedure from blood sample

By learning this process, students gain hands-on experience with fundamental molecular biology techniques used in research, medicine, and forensic science.

Objectives of the Experiment

1. To isolate genomic DNA from blood
2. To understand key steps: **cell lysis, purification, and DNA precipitation**
3. To practice safe handling of biological samples

Required Equipment

1. Microcentrifuge tubes (Eppendorf tubes)
2. Pipettes and tips
3. Centrifuge
4. Water bath/incubator

Required Reagents/Materials

1. Blood sample
2. RBC Lysis Buffer (RLB)

3. DNA Extraction Buffer
4. 10% SDS (Sodium dodecyl sulfate)
5. β-mercaptoethanol
6. Chloroform: Isoamyl alcohol (24:1)
7. Isopropanol (or ethanol)
8. 70% and 90% ethanol
9. TE buffer

Safety Precautions

1. Handle blood samples carefully (biohazard risk)
2. Wear gloves and lab coat
3. Work in a clean environment
4. Dispose of waste properly

Waste Disposal

1. Biohazard waste: Dispose of blood samples, tips, and tubes in biohazard containers.
2. Organic solvents: Collect chloroform mixtures in designated organic waste (do not pour in sink).
3. Alcohol waste: Place ethanol/isopropanol in flammable waste container.
4. Aqueous waste: Dispose of non-hazardous buffers in the designated aqueous waste container.
   Composition of Buffer Solutions
   a) **RBC lysis Buffer (RLB):** 0.155 mol/L NH4Cl, 10 mmol/L KHCO3 and 0.1 mol/L EDTA (Na2) in 1000 mL of distilled H2O. The pH should be adjusted to 7.6.
   b) **Extraction Buffer:** 1.5 mol/L Tris pH 7.6, 0.4 mol/L disodium salt of ethylenediaminetetra acetic acid (Na2EDTA; Merck, Darmstadt, Germany), 2.5 mol/L NaCl, 2% Cetyl trimethyl ammonium bromide (CTAB; Merck, Germany) 850 mL H2O. Adjust the pH to 8.0 and make the final volume to 1 L.
   c) **TE (Tris–EDTA) Buffer:** 10 mM Tris-HCl (pH ~8.0), 1 mM EDTA (Na_2EDTA) in 1000 ml of distilled water.

Experimental Procedure

1. Take ~500 μL of blood in a microcentrifuge tube. If frozen, thaw at room temperature first.
2. Centrifuge at 2664 RCF for 7 minutes at 4 °C. Carefully discard the liquid (plasma).

3. Add 1 ml RBC lysis buffer, mix gently, wait for 1-2 minutes. Centrifuge at 2664 RCF for 6 minutes at room temperature. Discard the supernatant. May be repeated this step until a white pellet (white blood cells) is obtained.
4. Add 500 µL prewarmed DNA extraction buffer, 30 µL 10% SDS, and 2 µL of β-mercaptoethanol respectively and mix gently. Incubate at 56–60°C for ~1 hour.
5. Add 500 µL chloroform:isoamyl alcohol, mix, and centrifuge at 10 656 RCF for 12 minutes at 4°C. Carefully collect the upper (clear) layer in a tube.
6. Add chilled ethanol to the tube. DNA will appear as white thread-like strands.
7. Centrifuge at 10656 RCF for 12 minutes at 4 °C to form a pellet and discard the liquid.
8. Wash with 90% (500 µL) and then 70% (500 µL) ethanol respectively using centrifugation (at 10656 RCF for 12 minutes at 4 °C) to remove impurities.
9. Dry the pellet at 37 °C and dissolve it in 100 µL TE buffer.
10. Store DNA at –20°C.

Observation/Comments

Post Lab Quiz

Experiment Title	
Date	
Name	
Lab Partner (s)	

1. Why are white blood cells (WBCs), and not red blood cells (RBCs), the primary source of genomic DNA in blood?

2. What is the role of RLB buffer in the DNA extraction protocol?

   a) To lyse WBCs
   b) To lyse RBCs
   c) To precipitate DNA
   d) To denature proteins

3. Why is ethanol used during DNA isolation, and what will happen if this step is omitted?

4. Name two downstream applications of the DNA isolated using this protocol.

5. If a student accidentally disturbs the DNA pellet while discarding the ethanol supernatant, what will be the consequence for the yield?

6. At what temperature should the final DNA solution be stored, and why is this temperature chosen?

Visualize Protein-Ligand Interactions with VMD

10

Introduction

VMD (Visual Molecular Dynamics) is a molecular visualization and analysis program designed to visualize biological systems such as proteins, nucleic acids, lipid bilayer assemblies, etc. The program is developed by the Theoretical and Computational Biophysics Group at the University of Illinois at Urbana-Champaign. Among molecular graphics programs, VMD is usually used for large biomolecular complexes and long-timescale molecular dynamics trajectories. It is also integrated with sequence information, as well as advanced image rendering and movie making to show biological system's interaction over time.

Registering VMD

VMD is free to use for students and researchers.
To download it:

- Go to the VMD website https://www.ks.uiuc.edu/Development/Download/download.cgi?UserID=&AccessCode=&ArchiveID=2
- Fill out the registration form
- Read and agree to the license. Once done, you can download the software.

Installing VMD

For Windows:

- Download the vmd installer file compatible for your windows computer.
- Double-click to run it
- Follow the on-screen instructions
- Make sure you have **administrator access**

For macOS:

- Download the vmd installer file compatible for your mac. Open, the downloaded VMD file (disk image)
- Drag the VMD app into your Applications folder
- VMD is ready to use (no extra steps needed)

Objective of the Experiment

1. Preparing a molecular interaction picture
2. Exploring the surrounding residues of a ligand bound with a protein
3. Picture rendering and saving the vmd state file

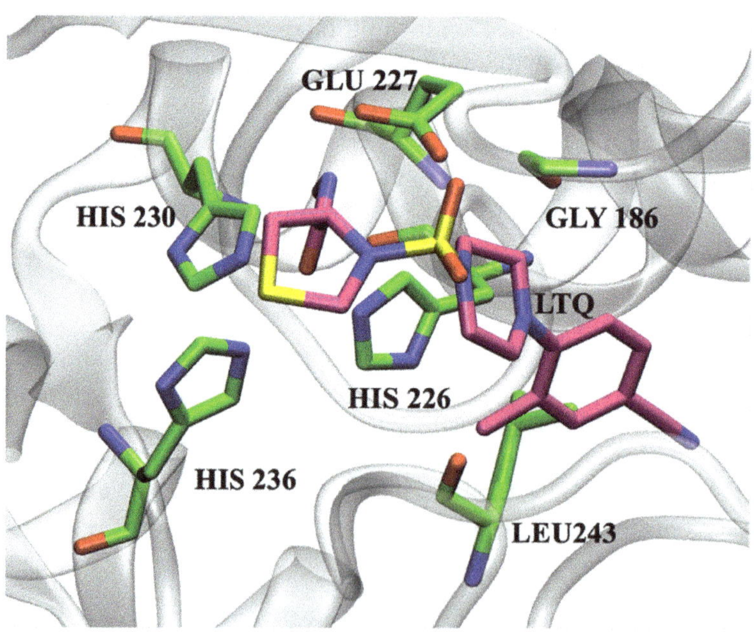

Figure: Representative molecular visualization that students will create using VMD

1. Preparing a molecular interaction picture

1.1 Load a pdb file using VMD

- Download the "Crystal structure MMP-9 complexes with a constrained hydroxamate based inhibitor LT4" (PDB ID: 5CUH) https://www.rcsb.org/structure/5CUH
- Click on the vmd icon and open it
- You will see three different windows: **VMD main**, **VMD open GL Display** and **Startup Command** windows

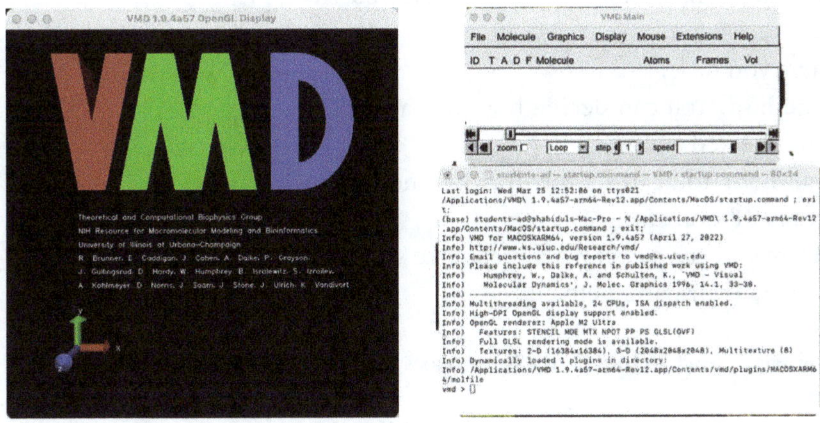

- You will use the VMD Main window and see the actions in VMD open GL Display window. You don't need the command window (You can minimize it)
- To load the pdb file, you can Drag and drop the pdb (5cuh.pdb) file on the Display.

or

- To load the pdb file follow: File → New Molecule → (Filename) Browse → Load

1.2 Change the Graphics/Display settings

- Graphics → Color → Display (Categories) → Background (Names) → White (Colors)
- Display → Orthographic
- Display → Uncheck Depth Cueing box

1.3 Protein and ligand presentation

By default, the molecular presentations are shown in line,

To present it as NewCartoon form for protein and licorice for ligand, you can follow the following steps:

Click on Graphics, then Representations. In the Graphical Representation window, you have the following important features that will help you to represent your molecule nicely.

- **Selected Molecule:** Here you will see the loaded
- Molecule (pdb file name)

- **Create Rep/Delete Rep:** You will use this to create/delete new representation of your molecule/selected specific portion of your molecule
- **Style/Color/Selection:** In this window all your selection will be shown in text. By default, the Style is Lines, Color is Name (according to name of the atoms) and the Selection is all
- **Selected Atoms:** You can select different portion of your system by using different keywords. You can know the keywords from the Selections button below the "Selected Atoms" window
- **"Draw style"** here you have:
    - **Coloring method:** You can decide how you are going to color your selected atoms.
    - **Drawing method:** You can set the drawing method as Lines/Cartoon/Ribbon etc.
    - **Material:** Here you can select the appearance style like Opaque, Glossy, Transparent etc.

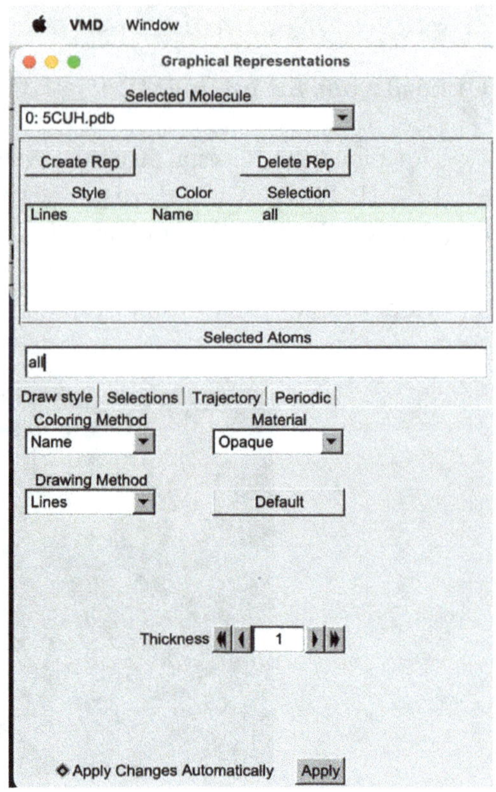

1.4 Presenting protein as "NewCartoon" and ligand as "licorice"

- You have two identical chains in this protein 5CUH. You need to focus on chain A. In the Selected Atoms box write: Chain A and hit Enter/Return button
- From Drawing Method select NewCartoon
- From "Coloring Method" select ColorID first and then the color number 8 (for white)
- Click on "Create Rep" it will create a new representation by copying the previous one. We will use it to present the ligand (LTQ).
- In the pdb file the ligand is not recorded as different chain here but it has a different residue name (resname is the keyword), which is LTQ
- Write "chain A and resname LTQ" hit enter
- From "Drawing Method" select "Licorice"
- We want to represent each atom type as different color for this ligand. Then from "Coloring Method" select Type. By default, the atoms will be colored as follows:
    - Carbon: Cyan, Nitrogen: Blue, Oxygen: Red, Sulfur: Yellow. In the reference picture we showed Carbon as Magenta. You can do this as follows:
    - From the VMD Main window click
    - Graphics → Colors → Type (Categories) → C (Names) → 28 Magenta2 (Colors)

2. Exploring the surrounding residues of a ligand bound with a protein

- Click "Create Rep" button to create a new representation and type "within 4 of chain A and resname LTQ" hit enter. It will show you different residues/residue fragments surroundings to the ligand (LTQ) molecule.

- From “VMD Main” window Click on Mouse → Label → Atoms 1
- Click on those residues/residue fragments on the VMD OpenGL Display. It will label the residues on the screen
- You can also find the labeled residue name and numbers from here, From the VMD Main window, Grphics → Labels. Note down the numbers (resid).
- You have found the surrounding residue numbers
- Double click on this representation to hide from the display
- Click “Create Rep” button to create a new representation and type “chain A and resid 226, 236, 230 …”
- Make sure the Drawing method is licorice.
- From Coloring Method select Name. By default, the colors are:
- Carbon: Cyan, Nitrogen: Blue, Oxygen: Red, Sulfur: Yellow. In the reference picture we showed Carbon as Lime. You can do this as follows, from the VMD Main window click
- Graphics → Colors → Name (Categories) → C (Names) → 12 Lime (Colors)
- In the reference picture, the protein is in transparent. Select the first representation click, Material → Transparent

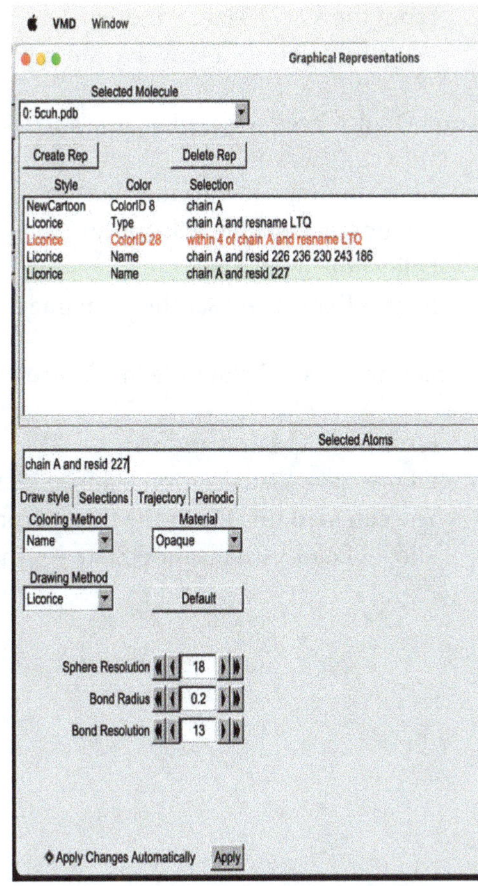

Note: On “vmd OpenGL Display” remove the labels as follows:

Graphics → labels (select all the labels showed in the display) → *Delete*

You also can zoom in, rotate or place your picture in a good position by using keyboard shortcuts or mouse/touchpad. The keyboard shortcuts are as follows:

R → to activate rotate mode

T → Translate mode

S → Scaling mode (zoom in and out)

Activate the required mode and use touchpad/mouse to apply this on the Display window

3. Picture rendering and saving the vmd state file

3.1 Picture rendering and save pictures

- You should remove the axis symbol from the window to have a clear picture by using Display → Axes → Off

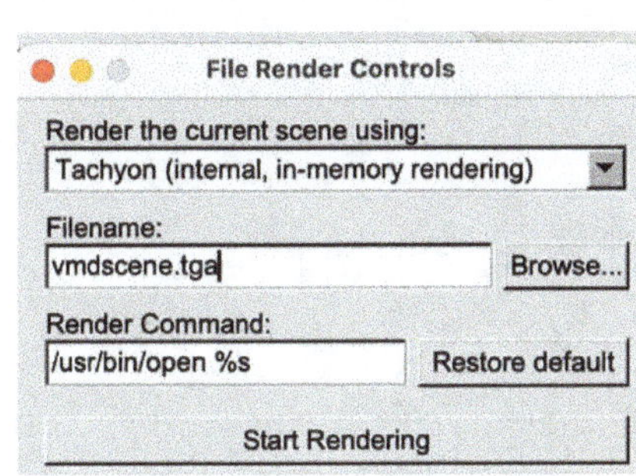

- From the VMD Main window:
  File → Render → Tachyon (internal, in-memory rendering)

Note: Do not need to set file name and the render command

- Click on "Start Rendering" button
- A vmdscene.tga file will appear. You need to export this as png file: Click on the vmdscene.tga file.
- Follow the followings:
  File → Export → (Set the file name in the Export as box) and save it where you want to keep the file.

3.2 Saving the vmd state file for future use

- From VMD Main window:
  File → Save Visualization State → (Set the name with. vmd extentesion)
- You can load this file in the future from the VMD Main window:
  File → Load Visualization State → The vmd state file. It will load exactly the current state.

Post Lab Quiz

Experiment Title	
Date	
Name	
Group Member (s)	

1. What is VMD? Name some uses of VMD in computational drug discovery studies.

2. How do you render and save the picture file in VMD? What is VMD state file?

3. How do you find the surroundings protein residues of a ligand bound with a protein?

4. Do the experiment with a protein-ligand complex pdb file (PDB ID: 6HYC, chain A). Find the interacting residues of the protein close to the ligand BH_4. Attach a picture that shows the residues within 6Å from BH_4.

Agarose Gel Electrophoresis for the Separation of DNA Fragments

11

Introduction

Agarose gel electrophoresis is a commonly used technique to separate DNA fragments based on their size. Agarose, a natural polymer obtained from seaweed, forms a porous gel when dissolved in buffer and cooled. These pores act like a sieve, allowing smaller DNA fragments to move faster than larger ones. DNA molecules carry a negative charge due to their phosphate backbone. When an electric field is applied, DNA migrates toward the positive electrode. Since DNA has a uniform charge-to-mass ratio, separation depends mainly on fragment size-smaller fragments travel farther through the gel. Several factors affect DNA migration, including agarose concentration, applied voltage, DNA size and shape, and buffer composition. After separation, DNA bands are visualized using a fluorescent dye under UV light.

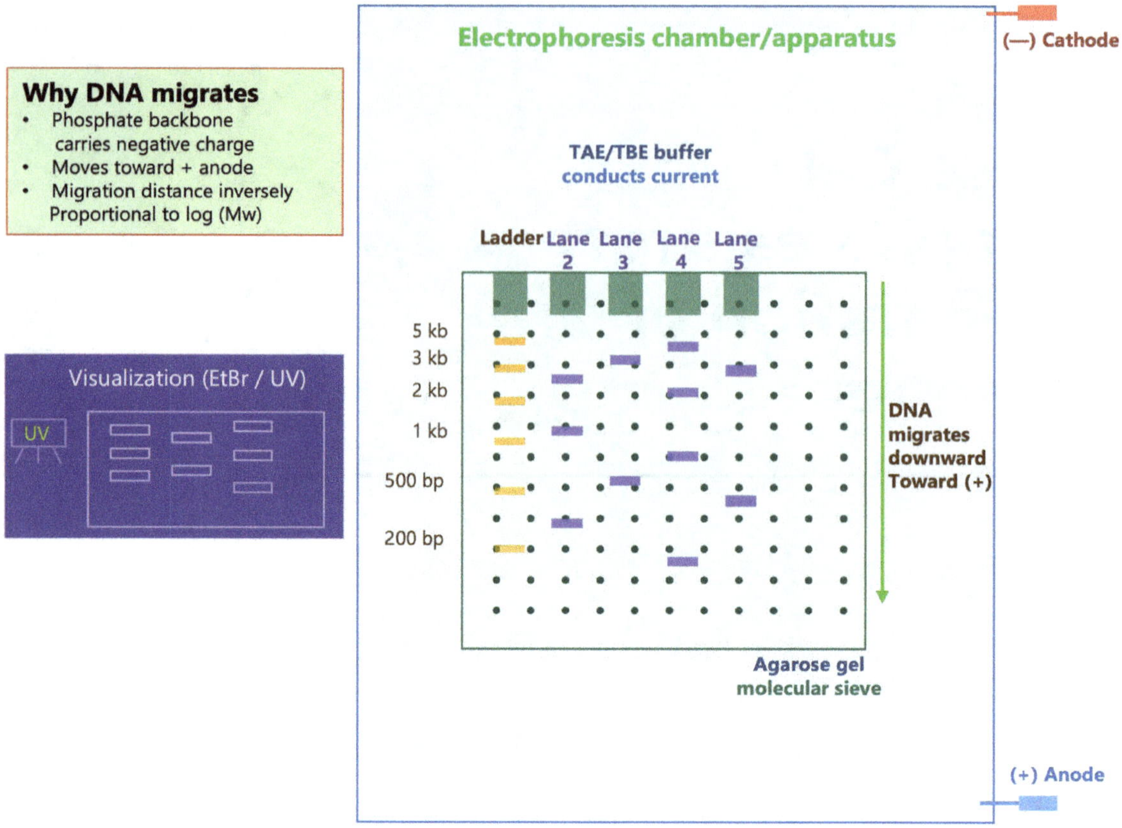

Figure: Agarose gel electrophoresis for separation of DNA fragments.

Objectives of the Experiment

1. To prepare and run an agarose gel and separate DNA fragments using agarose gel electrophoresis
2. To visualize DNA bands using a staining dye and estimate DNA fragment sizes using a DNA ladder

Required Equipment

1. Gel casting tray and comb
2. Electrophoresis chamber (gel box)
3. Power supply
4. Micropipettes and tips
5. Erlenmeyer flask
6. Microwave or hot plate
7. UV transilluminator / gel documentation system

Required Reagents/Materials

1. Agarose
2. Running buffer (TAE or TBE)
3. DNA samples
4. DNA ladder (size marker)
5. Loading dye
6. DNA staining dye (e.g., Ethidium bromide or alternative)

Safety Precautions

1. Always wear **gloves and lab coat**
2. Handle **DNA stains (e.g., EtBr)** with care (possible health hazard)
3. Avoid direct exposure to **UV light** (use protective shield/glasses)
4. Handle hot agarose carefully to prevent burns

Waste Disposal

1. Dispose of gels and buffers containing DNA stain in **designated chemical waste containers**
2. Do not pour stained solutions into the sink
3. Dispose of tips and consumables in **proper laboratory waste bins**
4. Follow institutional safety guidelines

Experimental Procedure

A. Preparation of Agarose Gel

1. Weigh 0.5 g of agarose (1% w/v) and add 50 ml buffer (1x TAE or 1x TBE) in a flask.
2. Heat the mixture until agarose fully dissolves (swirl occasionally). About 2-3 minutes might be enough.
3. Allow the solution to cool slightly (~60–65°C), then add 2.5 μL (1 drop) DNA stain.
4. Pour the gel into a casting tray with a comb and allow it to solidify for 20–30 minutes.

B. Setting Up and Running the Gel

1. Place the gel in the electrophoresis chamber and add the buffer solution (about 250-300 ml) to cover it.

2. Mix DNA samples with loading dye (15 µL DNA sample + 3 µL 6x loading dye).
3. Load 5–10 µL DNA ladder and 15-20 µL of sample into each well.
4. Connect the apparatus to the voltameter and run at 80 V for about 30 minutes.
5. Allow the dye to migrate to about two-thirds to three-quarters of the gel length.

C. Visualization of DNA

1. Turn off the power and carefully remove the gel.
2. Place the gel on a UV transilluminator.
3. Observe and record the **fluorescent DNA bands**.

Observation / Comments

Post Lab Quiz

Experiment Title	
Date	
Name	
Lab Partner (s)	

1. Why do DNA fragments migrate toward the positive electrode (anode) during agarose gel electrophoresis?

2. How does agarose concentration affect the pore size of the gel?

3. Describe the relationship between DNA fragment size and migration distance in an agarose gel.

4. What is the role of the loading buffer when preparing samples for gel electrophoresis?

5. If you applied too high voltage during electrophoresis, what might you observe?

Understanding Genetic Codons, Mutations, and the Prediction of Deleterious Mutations on Protein Structure and Function Using Computational Tools

12

Introduction

The genetic code is the system by which DNA sequences are translated into proteins and follows the **central dogma of molecular biology (DNA → RNA → Protein)**. DNA contains four nucleotides: **adenine (A), thymine (T), cytosine (C), and guanine (G),** and this information is first transcribed into messenger RNA (mRNA), where thymine (T) is replaced by **uracil (U)**. There are four nucleotides, and a minimum of three nucleotides is required to encode all 20 amino acids. These three-nucleotide sequences are called codons, and together they form a triplet code ($4^3 = 64$ possible codons). In this experiment, the codons are interpreted using an **mRNA codon table,** which shows how each codon corresponds to a specific amino acid.

The genetic code is nearly universal, meaning most organisms use the same codons to specify amino acids. It is also redundant, as multiple codons can encode the same amino acid. By reading an mRNA sequence using the codon table, the corresponding protein sequence can be determined.

The Standard Genetic Code — mRNA Codons

First Base of Codon	Second Base of Codon (5'→3' reading)								Third Base of Codon
	codon U	amino acid	codon C	amino acid	codon A	amino acid	codon G	amino acid	
U	UUU	*phe*	UCU	*ser*	UAU	*tyr*	UGU	*cys*	U
	UUC		UCC		UAC		UGC		C
	UUA	*leu*	UCA		UAA	STOP	UGA	STOP	A
	UUG		UCG		UAG	STOP	UGG	*trp*	G
C	CUU	*leu*	CCU	*pro*	CAU	*his*	CGU	*arg*	U
	CUC		CCC		CAC		CGC		C
	CUA		CCA		CAA	*gln*	CGA		A
	CUG		CCG		CAG		CGG		G
A	AUU	*ile*	ACU	*thr*	AAU	*asn*	AGU	*ser*	U
	AUC		ACC		AAC		AGC		C
	AUA		ACA		AAA	*lys*	AGA	*arg*	A
	AUG	met / START	ACG		AAG		AGG		G
G	GUU	*val*	GCU	*ala*	GAU	*asp*	GGU	*gly*	U
	GUC		GCC		GAC		GGC		C
	GUA		GCA		GAA	*glu*	GGA		A
	GUG		GCG		GAG		GGG		G

STOP codon   START (AUG = met)

Figure: Standard genetic code (mRNA) table

Mutations are changes in the DNA sequence that may alter the corresponding mRNA codons and, consequently, the amino acid sequence of a protein. These changes can affect protein structure, stability, and function. Some mutations are harmless, while others may be harmful (deleterious).

Figure: Different types of mutations

In this experiment, computational tools such as SIFT and **PolyPhen-2** are used to predict the effect of amino acid substitutions on protein function. Human insulin, a small protein involved in glucose regulation, is used as the model system.

Objectives of the Experiment

1. To understand the genetic codon system and translation of DNA into proteins
2. To learn how mutations affect protein structure and function
3. To analyze amino acid substitutions using computational tools
4. To classify mutations as **benign or deleterious**

Required Equipment

1. Computer with internet access
2. Web browser

Required Reagents/Materials

1. Protein sequence (FASTA format) of insulin
2. Access to online tools:
    - SIFT
    - PolyPhen-2

Safety Precautions

1. Ensure proper use of computers and internet resources
2. Do not share personal information unnecessarily
3. Follow academic integrity while recording results

Waste Disposal

1. No chemical or biological waste is generated
2. Dispose of printed materials in regular waste bins
3. Save digital data responsibly

Experimental Procedure

Part A: Understanding Genetic Codons

1. Study the codon table provided by the instructor.
2. Practice identifying amino acids from given codons.
3. Understand redundancy and universality of the genetic code.

Part B: Prediction of Deleterious Mutations

Test Protein Sequence (Insulin)
FASTA FORMAT:

MALWMRLLPLLALLALWGPDPAAAFVNQHLCGSHLVEALYLVCGERGFFYTPKTRREAEDLQVGQ-VELGGGPGAGSLQPLALEGSLQKRGIVEQCCTSICSLYQLENYCN

Using SIFT Tool

1. Open: https://sift.bii.a-star.edu.sg/www/SIFT_seq_submit2.html
2. Paste the insulin sequence in FASTA format
3. Enter mutation(s) (e.g., A12G, W17F)
4. Use default settings
5. Submit the query
6. View results and record the SIFT score

Interpretation:

- Score < 0.05 → Likely deleterious
- Score ≥ 0.05 → Likely tolerated

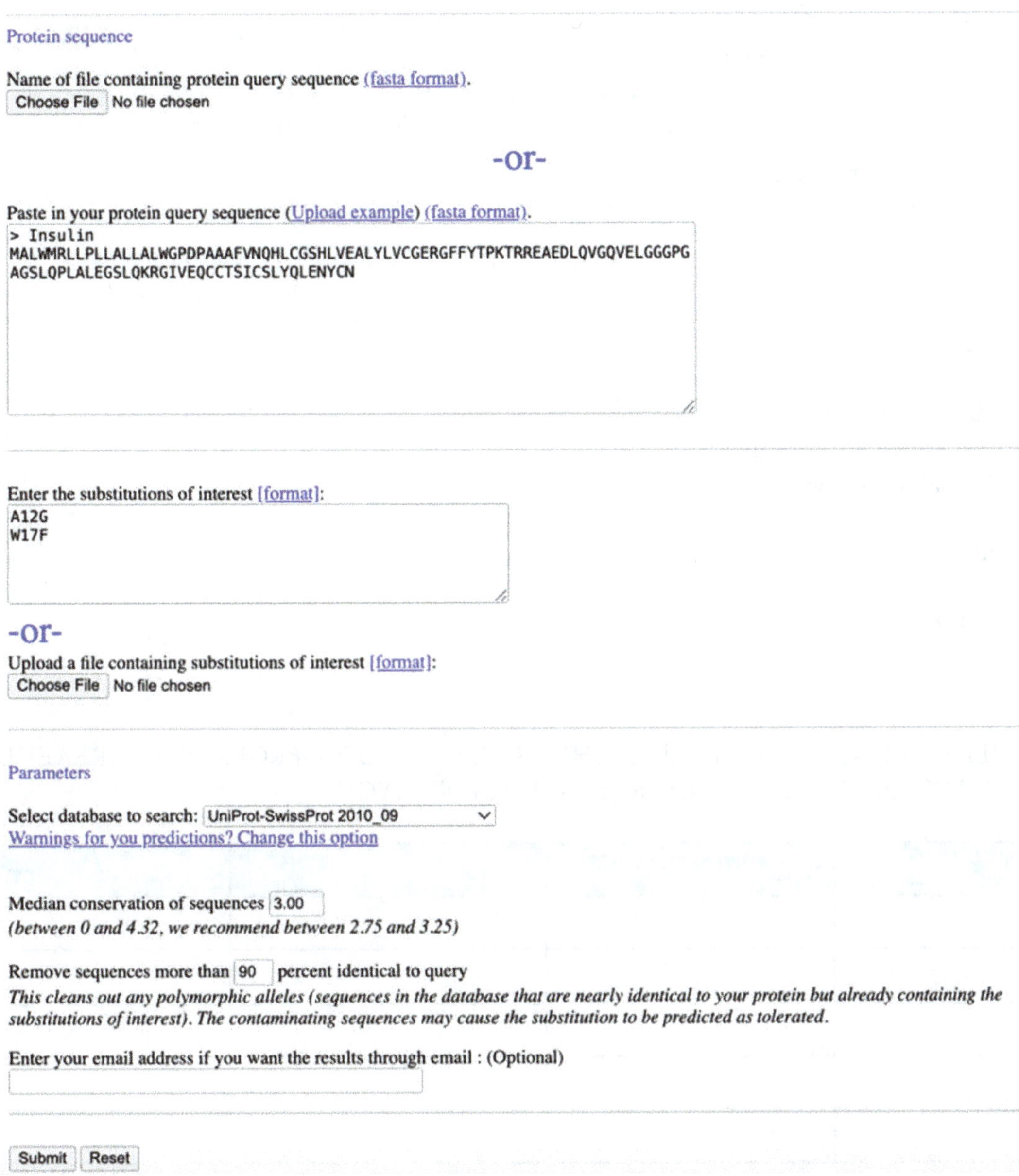

Using PolyPhen-2 Tool

1. Open: http://genetics.bwh.harvard.edu/pph2/
2. Input the protein sequence
3. Enter mutation details (position and amino acid change)
4. Submit the query
5. View and record results

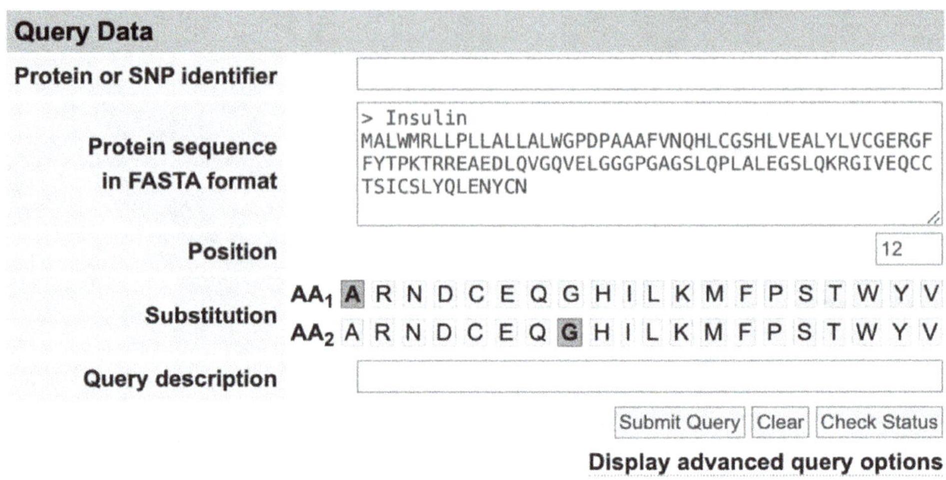

Interpretation:

- Score < 0.15 → Benign
- 0.15–0.85 → Possibly damaging
- 0.85 → Probably damaging

Observation:

Studied protein

> Insulin

MALWMRLLPLLALLALWGPDPAAAFVNQHLCGSHLVEALYLVCGERGFFYTPKTRREAEDLQVGQ-VELGGGPGAGSLQPLALEGSLQKRGIVEQCCTSICSLYQLENYCN

Mutations	SIFT score	Polyphen-2 score	Comments
A24D			
H29D			
G32R			
L35P			
C43G			

Post Lab Quiz

Title of Experiment	
Date of Experiment	
Name	
Lab Partner (s)	

1. A trinucleotide-based code would provide 4^3 or 64 codons. Are the extra codons used?

2. The codon is _______ (choose from 1, 2, 3, 4) bases long and is part of an _______ (choose from mRNA, rRNA, tRNA)

3. What is the codon for methionine?

4. What are the significances of the term START and STOP in the Table?

5. Which terminus of a newly forming polypeptide chain contains the first (initiating) amino acid—the N-terminus or the C-terminus?

6. Is there a codon on mRNA that functions as a signal to terminate polypeptide chain elongation?

7. The following represents the full mRNA sequence from a bacterial cell:

5' ACUAGCAGGAGACGUAAAUGGCGAUGUGCCAGAUGCGCAGUCACACAUAACUGCAAG 3'

a. How many amino acid residues make up the protein?

b. What is the total number of tRNAs that participate in the translation of this protein?

c. Determine the amino acid at position 2 of the polypeptide chain.

8. Convert the following mRNA codons into the corresponding DNA sequence: arginine (CGU), leucine (UUA), valine (GUC).

a. CGUUUAGUC
b. GCAAAUCAG
c. GCAAATCAG
d. CGTTTUGAC

9. A single-chain protein contains a mutation in which the N-terminal phenylalanine is replaced by leucine. Which level(s) of protein structure does this change affect? (Circle the best answer.) Assume no quaternary structure.

a. primary only
b. primary and secondary
c. primary, secondary and tertiary
d. primary, secondary, tertiary and quaternary

10. Which of the following DNA sequences will best maintain its double-stranded structure at higher temperatures? (Only one strand is shown.) Circle the correct answer.

a. CATCCTAGCGACTAT
b. CTATACGACATAGCC
c. AAATGCATACATCTT
d. CCCGCATCGCCATCG

11. Which of the following **BEST** depicts the concept of a mutation?

a. an abnormal base sequence existing at birth
b. a change in the base sequence that causes serious injury or death
c. any alteration in the base sequence
d. an alteration in the base sequence of the gametes

12. What are the different causes of genetic mutations?

13. List different types of genetic mutation.

14. How can a mutation disrupt the structure and function of a protein?

15. Which of the studied mutations can affect the insulin protein most deleteriously based on SIFT and Polyphen2 prediction?

References

(1) Vincent, J. B.; Crowder, M. W.; Averill, B. A. Evidence for a Phosphoryl-Enzyme Intermediate in Phosphate Ester Hydrolysis by Purple Acid Phosphatase from Bovine Spleen. *Journal of Biological Chemistry* **1991**, *266* (27), 17737–17740. https://doi.org/10.1016/S0021-9258(18)55186-3.

(2) Aich, S.; Delbaere, L. T. J.; Chen, R. Continuous Spectrophotometric Assay for β-Glucuronidase. *Biotechniques* **2001**, *30* (4), 846–850. https://doi.org/10.2144/01304RR02.

(3) Seibert, E.; Tracy, T. S. Fundamentals of Enzyme Kinetics: Michaelis-Menten and Non-Michaelis-Type (Atypical) Enzyme Kinetics. *Methods Mol. Biol.* **2021**, *2342*, 3–27. https://doi.org/10.1007/978-1-0716-1554-6_1.

(4) Guha, P.; Das, A.; Dutta, S.; Chaudhuri, T. K. A Rapid and Efficient DNA Extraction Protocol from Fresh and Frozen Human Blood Samples. *J. Clin. Lab. Anal.* **2017**, *32* (1), e22181. https://doi.org/10.1002/JCLA.22181.

(5) Humphrey, W.; Dalke, A.; Schulten, K. VMD: Visual Molecular Dynamics. *J. Mol. Graph.* **1996**, *14* (1), 33–38. https://doi.org/10.1016/0263-7855(96)00018-5.

(6) Lee, P. Y.; Costumbrado, J.; Hsu, C. Y.; Kim, Y. H. Agarose Gel Electrophoresis for the Separation of DNA Fragments. *J. Vis. Exp.* **2012**, No. 62. https://doi.org/10.3791/3923.

(7) *Amino Acid Translation Table.* https://web.archive.org/web/20200529000711/http://sites.science.oregonstate.edu/genbio/otherresources/aminoacidtranslation.htm (accessed 2026-03-30).

(8) Clancy, S. Genetic Mutation. *Nature Education* . 2008, p 187.

www.ingramcontent.com/pod-product-compliance
Lightning Source LLC
LaVergne TN
LVHW082356200926
848481LV00012B/3579